W9-CIC-686

EARTH IN ACTION

HURRICANES

by Joanne Mattern

Content Consultant
Stephen A. Nelson
Associate Professor of Geology
Tulane University

Published by ABDO Publishing Company, PO Box 398166, Minneapolis, MN 55439.

Printed in the United States of America,
North Mankato, Minnesota
052013
092013
THIS BOOK CONTAINS AT LEAST 10% RECYCLED MATERIALS.

Editor: Lauren Coss
Series Designer: Becky Daum

Library of Congress Control Number: 2013932181

Cataloging-in-Publication Data
Mattern, Joanne.
Hurricanes / Joanne Mattern.
p. cm. -- (Earth in action)
ISBN 978-1-61783-939-9 (lib. bdg.)
ISBN 978-1-62403-004-8 (pbk.)
1. Hurricanes--Juvenile literature. 2. Natural disasters--Juvenile literature. I. Title.
551.55--dc23

2013932181

Photo Credits: Dr. Robert Muntefering/Getty Images, cover, 1; Gerry Broome/AP Images, 4, 39; Red Line Editorial, 6, 16; Will Waldron/Times Union/AP Images, 8; Glynnis Jones/Shutterstock Images, 10, 19, 45; NASA/AP Images, 12, 23; Thinkstock, 14, 36; Jose Goitia/AP Images, 20; El Nuevo Diario/AP Images, 25; Dave Martin/AP Images, 26; J Pat Carter/AP Images, 28; AP Images, 31; Stephan Savoia/AP Images, 34; Jose Luis Magana/AP Images, 40

CONTENTS

CHAPTER ONE

SUPERSTORM!

On October 22, 2012, meteorologists were busy studying their weather maps. They saw a tropical depression had formed just south of Jamaica. A tropical depression is an area of strong thunderstorms spinning around an area of low pressure. The scientists kept an eye on the storm. They knew the depression could spell trouble.

Waves from Hurricane Sandy pound a pier in North Carolina on October 27, 2012. Meteorologists had been tracking the storm for days before it hit.

How Hurricanes Get Their Names

In 1954 meteorologists started giving each hurricane a different woman's name. The list started with the letter *A* and went all the way to the letter *Z*. In 1979 the list was changed to use both men's and women's names. Meteorologists now use six name lists, starting at the beginning of a new list each year. If a storm is particularly destructive, its name is retired and never used again.

Sandy Strikes

Sure enough, the tropical depression got stronger. It became a tropical storm. The storm moved north and east. Soon it was a hurricane that rated a Category 2 on the Saffir-Simpson hurricane wind scale. The scientists named the storm Sandy. They watched as Sandy moved north across the Atlantic Ocean.

On October 23, the storm began crossing the islands of Jamaica, Cuba, and the Bahamas. More than 20 inches (51 cm) of rain fell in Haiti in 24 hours. The excess water flooded rivers and streets. Crops and houses were destroyed in Jamaica and Cuba.

Category	Wind Speed	Damage
1	74–95 mph (119–153 km/h)	Some damage to homes and tree branches; power outages likely
2	96–110 mph (154–177 km/h)	Extensive damage to homes, including roofs and siding; many trees blown down; widespread power outages
3	111–129 mph (178–208 km/h)	Devastating damage to homes, including many roofs removed; widespread damage to trees; power outages may last for weeks
4	130–156 mph (209–251 km/h)	Catastrophic damage to many homes; most trees will be blow down; the area may not be livable for weeks to months
5	157 mph (252 km/h) or higher	Catastrophic damage to most homes, including collapsed roofs and walls; most of the area will not be livable for weeks to months

The Saffir-Simpson Scale

This chart explains how meteorologists classify hurricanes using the Saffir-Simpson scale. How does the information in this chart compare to what you have learned from the text about the strength and effects of hurricanes? How does this chart help people understand what to expect from an approaching hurricane?

Meteorologists track Hurricane Sandy on October 30.

The hurricane killed more than 70 people on its path through the Caribbean.

By October 28, Sandy was a huge storm. Eventually it would reach 1,100 miles (1,770 km) in diameter, making Sandy the largest Atlantic hurricane ever recorded. Warnings went out to the residents of the northeastern United States.

Sandy Hits the East Coast

Most Atlantic hurricanes move northwest, but the cold front blocked Sandy from doing that. Instead the hurricane turned left and made landfall in Atlantic

City, New Jersey, on October 29. The storm continued moving northwest. It spread destruction from New Jersey all the way up through New England.

Most people in the area had never seen anything like this storm. Some areas in New Jersey had gusts over 90 miles per hour (145 km/h). A powerful storm surge pushed the ocean onto the land with tremendous force. Floodwaters up to eight feet (2.4 m) deep raced up streets and into houses. People were unable to return to their homes for months. Entire communities in New Jersey and New York were destroyed.

Storm Surges

A storm surge occurs when a hurricane's winds push seawater ahead of the storm. This can raise sea level several feet above normal, causing flooding that can wipe away coastal communities. Storm surges are most common along the coast. However, they can also extend into other bodies of water. The storm surge from Hurricane Sandy drove water over the banks of the Hudson River, the Long Island Sound, and other bodies of water, causing terrible flooding in nearby communities.

Sandy's storm surge swept the roller coaster in Seaside Heights, New Jersey, out to sea.

At least 24 states felt Sandy's effects. The storm's influence stretched as far north as Maine and as far west as Michigan and Wisconsin. A ship sank off the coast of North Carolina. High winds knocked down trees and power lines across the eastern United States. Heavy snow fell in the mountains of West Virginia when Sandy's moisture slammed into the freezing temperatures, changing rain to snow. The storm killed more than 110 people in the United States.

STRAIGHT TO THE SOURCE

Alexander Hamilton grew up to be an important figure in the American Revolution (1775–1783). But in 1772 he was a teenager living in Saint Croix in the Caribbean's Virgin Islands. He wrote a letter to his father after a violent hurricane struck Saint Croix:

> *I take up my pen just to give you an imperfect account of the most dreadful hurricane that memory or any records whatever can trace. . . . The roaring of the sea and wind . . . the prodigious glare of almost perpetual lightning—the crash of the falling houses—and the ear-piercing shrieks of the distressed were sufficient to strike astonishment into Angels. A great part of the buildings throughout the Island are levelled to the ground—almost all the rest very much shattered—several persons killed and numbers utterly ruined.*

Source: Alexander Hamilton. "St. Croix, September 6, 1772." A Few of Hamilton's Letters: Including His Description of the Great West Indian Hurricane of 1772. *Gertrude Atherton, ed. New York: Macmillan Company, 1903. Print. 261–262.*

Consider Your Audience

This letter was written for an audience living in the 1700s. How might this letter be different if it was written today? Write a blog post conveying the information in this letter to a modern audience. How is your blog post different from the original letter?

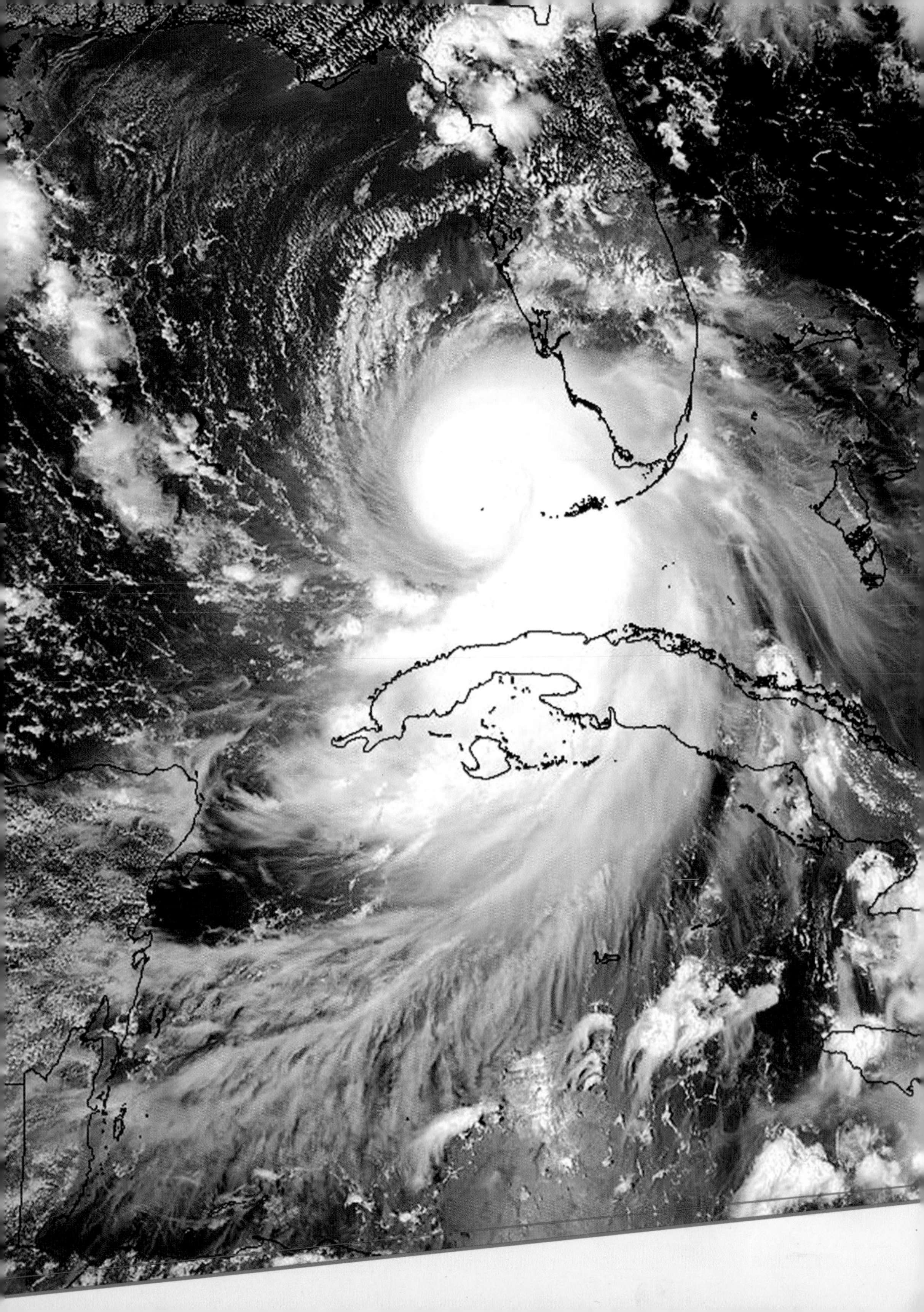

CHAPTER TWO

A HURRICANE IS BORN

The hottest places on Earth tend to be around the middle of the planet, near the equator. The coldest places on Earth are at the North and South Poles. The atmosphere and the oceans are always trying to make these temperatures equal. Hot air masses and warm ocean currents move from the tropics toward the poles. At the same time, colder air and ocean currents move from the poles toward the

It takes a very special set of climate events to create a hurricane. These events determine where the storm will begin, how strong it will be, and what path it will take.

As warm air rises, the cold air takes its place, eventually creating a rotating storm system.

tropics. Low pressure occurs when warm air moves upward. High pressure occurs when cool air moves downward. Air moves from areas of high pressure to areas of low pressure. This movement creates wind. The larger the difference in pressure, the stronger the wind.

Tropical and polar air masses and ocean currents often meet in the tropics. When warm seas meet cooler air, violent storms can develop. When tropical seas reach a temperature of 80 degrees Fahrenheit (27°C) or more, a current of warm, wet air rises. This creates an area of low pressure. Cold air is heavier than warm air. Cold air will always rush in to fill the space left by the rising warmer, lighter air. As the cold winds rush in, the Earth's rotation causes the winds to spin.

In the Northern Hemisphere, these winds spin in a counterclockwise direction. In the Southern Hemisphere, they spin clockwise. When a spinning area of low pressure and high winds reaches 35 miles per hour (56 km/h), it becomes a tropical storm. In the Atlantic Ocean, hurricane season lasts from June 1 to November 30. Most storms form in late summer through early fall, when the ocean water is warmest.

Eastern Pacific Hurricanes

The Eastern Pacific hurricane season runs from May 15 to November 30. On average, 18 tropical storms form each year over the Eastern Pacific. Every so often, one of these storms will develop into a hurricane. These storms typically hit Mexico's west coast or Hawaii. Because of cooler ocean waters and typical wind direction, few hurricanes have hit the US West Coast.

From Tropical Storm to Hurricane

Tropical storms can bring a lot of rain and high winds. Sometimes the storms fade away after a few

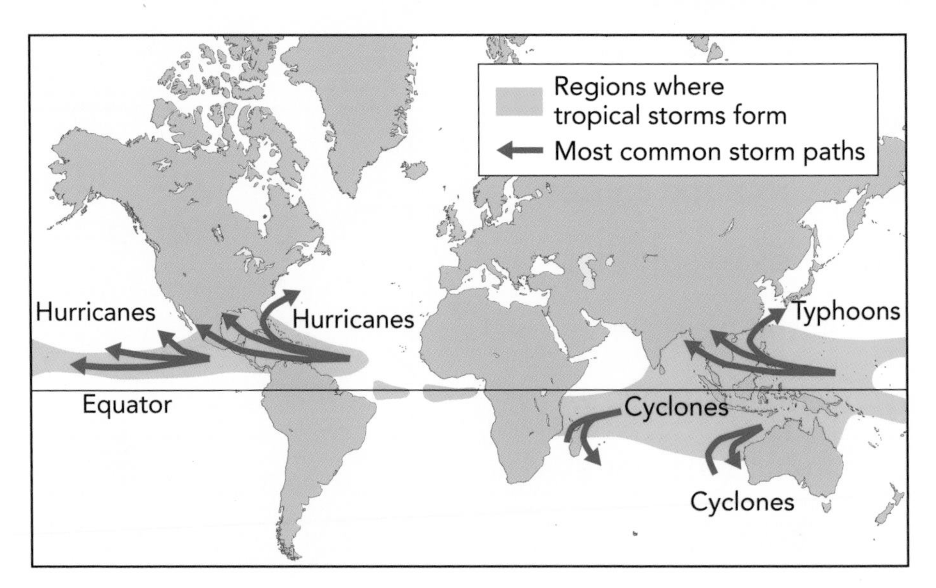

Where Hurricanes Occur
This diagram shows where hurricanes and other tropical storms occur around the world and the direction of the storms' typical paths. After looking at the map, what do you notice about the areas where these storms strike? How does seeing this map help you understand the weather conditions that can cause hurricanes to form?

days. If the winds continue to grow, a tropical storm might turn into a hurricane. Once wind speeds reach 74 miles an hour (119 km/h), the storm is officially called a hurricane. The same type of storm is called a typhoon in the western Pacific Ocean. In the Indian Ocean, this storm is called a cyclone. No matter what the name, it is the same kind of storm with the same conditions.

Hurricanes are huge storms. They can have a diameter of 400 to 500 miles (640–800 km). Hurricanes always form at sea. But they don't always stay there. Atlantic hurricanes usually move westward because of the movement of the wind. Hurricanes move about 10 to 20 miles per hour (16–32 km/h). However, hurricanes sometimes slow down or speed up.

A hurricane gains much of its energy from the warm water. As the storm moves over land, it loses the energy from the warm water. A hurricane usually gets weaker over land. It might turn

Watches and Warnings

If a storm is approaching, the National Weather Service issues watches and warnings. A watch alerts people to weather conditions likely to produce a storm. Watches tell people to be on guard and stay informed about weather conditions. Watches are serious, but they are not as serious as warnings. A storm warning is issued when meteorologists are relatively confident a storm will hit a given area. Warnings tell people to take immediate action, such as finding shelter or evacuating.

back into a tropical storm. However, even these less powerful storms can still bring a lot of heavy rain and high winds.

The National Oceanic and Atmospheric Administration (NOAA) is a government agency that tracks weather conditions. According to NOAA, an average of ten tropical storms form in the Atlantic Ocean, Caribbean Sea, or Gulf of Mexico every year. About six of these storms become hurricanes. Most hurricanes stay out at sea. However, about five hurricanes strike the United States every three years.

Of the five hurricanes that strike the United States every three years on average, two are major hurricanes that do serious damage, such as Hurricane Sandy.

INSIDE A HURRICANE

Several special features make a storm a hurricane. These features explain why a hurricane acts the way it does. They also explain what kind of damage a hurricane can cause.

Eye of the Storm

A major feature of a hurricane is its eye. As the hurricane's winds rotate, they create a calm area in the middle of the circle of clouds. This eye is usually about

Because they are such large storms, hurricanes can affect a large area for several hours or even days.

Hurricane Emergency Kit

Hurricanes are very dangerous storms. It is important to know what to do if a hurricane is headed your way. If you live in a hurricane-prone area, it's a good idea to have an emergency kit ready before the storm strikes. Your kit should include:

- flashlights
- battery-powered or hand-cranked radio
- extra batteries
- water
- canned food and a can opener
- first-aid supplies
- medicine
- blankets
- cell phone

20 to 40 miles (32–64 km) in diameter. Conditions inside the eye are not stormy at all. The winds in the eye are light, and not much rain falls. The sky may even be clear, with stars showing or the sun shining.

The thick clouds that surround the eye are called the eye wall. The eye wall is the most dangerous part of the storm. Conditions inside the eye wall include severe thunderstorms, high winds, and heavy rain.

Violent weather can stretch out from the eye of a hurricane for several hundred miles. Conditions are better farther away from the eye.

Lots of Water

Heavy rain is the most significant feature of a hurricane. Hurricanes can produce amazing rainfall totals. Some hurricanes can produce six inches (15 cm) of rainfall in one hour.

Heavy rain causes extreme flooding. Floods are the most dangerous feature of a hurricane. About 90 percent of the deaths that occur during hurricanes result from people drowning in floods. Floods are common in coastal areas when powerful ocean waves crash onshore with tremendous force. Storm surges also contribute to heavy flooding.

Hurricane flooding is not limited to coastal areas. Sudden floods, called flash floods, can occur anywhere that heavy rain causes rivers and streams to overflow. During the Category 3 Hurricane Irene in 2011, inland streams and rivers overflowed in Vermont and New York. This caused severe flood damage in towns hundreds of miles from the ocean.

Environmental Impact

Hurricanes don't just affect people and property. They also affect nature itself. Storm surges erode beaches and coastal land. Landslides and floods can wash away rocks, trees, and other plants and leave an area bare of vegetation. High winds can also destroy acres of trees. Hurricane Sandy knocked down more than 8,000 trees in New York City alone. The storm also demolished thousands of trees in parks and wilderness areas, changing the landscape for years to come.

The landslides from 1998's Hurricane Mitch destroyed more than 25 villages in Nicaragua.

A hurricane can also trigger other disasters. In Central America, where there are many high mountains, landslides are common results of heavy rain and flooding. In 1998 the Category 5 Hurricane Mitch dumped more than ten inches (25 cm) of rain in Honduras and Nicaragua, which caused many landslides.

Key West, Florida, residents battle high winds during Hurricane Georges in 1998.

Hurricane Winds

Water is not the only dangerous part of a hurricane. Thunderstorms are often associated with hurricanes. Some hurricanes even develop tornadoes. And all hurricanes have high winds. These winds create a lot of damage and danger.

High winds knock down power lines and uproot trees. These winds can rip roofs off buildings and flip

EXPLORE ONLINE

Chapter Three describes what goes on inside a hurricane and the effects of hurricanes. The link at the Web site below also gives information about the dangers that can result from hurricanes and how to prepare for them. As you know, every source is different. How is the information on this Web site the same as information in the chapter? How is it different? What can you learn from this Web site?

Hurricanes

www.mycorelibrary.com/hurricanes

over cars. Debris blowing through the streets can become deadly weapons. Falling trees and downed power lines kill many people during hurricanes. They can leave people without power for weeks after the storm has ended.

CHAPTER FOUR

STORM TRACKERS

In the past, people looked at rising waves and thick storm clouds in the sky. They felt the rising wind. But they did not get much advance warning of storms headed their way. Often the people living in an area struck by a storm had no time to escape. People had no way of knowing a hurricane was coming until the storm was nearby.

A National Hurricane Center (NHC) meteorologist studies Hurricane Irene in 2011.

The Galveston Hurricane of 1900

On September 8, 1900, the mightiest storm in US history ripped through Galveston, Texas. Almost every structure on Galveston Island was destroyed by the 15-foot (4.6-m) storm surge and 120-mile-per-hour (193-km/h) winds. When the wind and rain ended, more than 6,000 people were dead.

Deadly Mistakes

During the 1920s and 1930s, the National Weather Service did its best to warn people when dangerous storms were coming. However, it often made mistakes. In 1935 meteorologists started tracking a storm near Florida. At first they thought the storm would miss the southern tip of the state. However, the storm struck the Florida Keys with tremendous force.

Better Predictions

Fortunately, technology improved during the middle of the 1900s. During the 1950s and 1960s, meteorologists began using computers and satellites to track hurricanes. Weather forecasters could see where storms were headed. They could figure out

The Galveston Hurricane of 1900 devastated the city of Galveston, leaving more than 10,000 people homeless.

how strong a storm was. They could issue warnings to people in the path of a hurricane. People had time to prepare or to evacuate. These warnings have saved many lives.

Today US hurricane forecasts come from the National Hurricane Center (NHC). The NHC is part of the National Weather Service. It is located in Miami, Florida. NHC scientists monitor satellites, radio broadcasts, and wind gauges. They gather data and enter it into computers. Within seconds they can see exactly where a storm is and make predictions about where it might be headed.

Hurricane trackers also rely on people who fly directly into the storm. These hurricane hunters use scientific instruments to measure conditions inside a storm. These conditions include the temperature, wind speed, barometric pressure, and size of the storm. This important information has helped save many lives.

Hurricane Trackers

Flying directly into the center of such a violent storm can be very dangerous. But it also provides valuable information that scientists cannot get anywhere else. That information can save lives. In 2005 a remote-controlled plane flew into a hurricane for the first time. Scientists are still working to perfect the technology. But someday we may be able to gather data from the inside of a storm without putting human lives in danger.

STRAIGHT TO THE SOURCE

In 1989 meteorologist Jeffrey M. Masters was part of a mission to fly into the eye of Category 5 Hurricane Hugo. He later described his experience:

> *We hit the eye wall. Darkness falls. Intense blasts of turbulent wind rock the airplane. Torrential rain hammers the fuselage. The winds shoot up to 170 mph, gusting to 190. The three remaining engines whine and roar as [the pilot] fights off a powerful updraft. The turbulence is rough, but survivable. We cross the inner eye wall without hitting any incredible jolts like those that nearly knocked us from the sky on our way in. . . . One minute gone, half-a-minute to go. The airplane is barely shaking now, the turbulence is so light. It is hard to believe we are in the eye wall of Hugo!*

Source: Jeffrey M. Masters. "Hunting Hugo."
Weatherwise *September 1999: 20. Web. Accessed March 26, 2013.*

What's the Big Idea?

What is the main point Masters is trying to convey about his experience in Hugo? Choose two sentences from the passage that back up your answer. Why do you think Masters chose the words he did to describe his experience?

CHAPTER FIVE

TOMORROW'S HURRICANES

Even though people have studied hurricanes for many years, these powerful storms can still surprise us. Every year scientists try to predict how many hurricanes will strike. Some years have more storms than others.

Climate Change

In 2005 the Atlantic had one of the most active hurricane seasons since record keeping began.

Meteorologists do their best to study hurricanes. However, these storms can be hard to predict.

Hurricane Katrina was one of the many hurricanes of 2005. It caused extensive flooding in New Orleans, Louisiana.

NOAA put out a report trying to explain why there had been so many hurricanes. The report stated that the hurricanes were part of a normal cycle of weather conditions. However, many people believe there is another reason hurricanes have gotten more powerful and more numerous. They think the answer is climate change. Most scientists believe that people and industries have caused an increase in the atmosphere's temperature. Earth's average temperature is getting warmer. This increase doesn't mean that every area on Earth is always warmer. However, it does mean that ocean temperatures are rising. Those higher temperatures more often create the perfect conditions for a hurricane to form.

In addition, global warming is melting the ice trapped at Earth's North and South Poles. This melting puts so much water into the oceans that it raises the sea level. The oceans may only rise by a few inches. However, that is enough to put large coastal areas in danger

El Niño and La Niña

As water moves around the world's oceans, it has a big effect on the weather. Two currents are especially important. El Niño is a warm ocean current that appears near the northwest coast of Peru. Every two to seven years, El Niño appears much farther south than usual. This causes ocean temperatures to rise. That change in ocean temperature affects weather around the world. A major El Niño can last three or four years. El Niño is usually followed by an opposite event, called La Niña. During La Niña, ocean temperatures are colder than normal. Both El Niño and La Niña cause extreme weather, including dangerous hurricanes.

Coastal Construction

In the past, a lot of coastal land was undeveloped. When a strong storm struck, there weren't many houses and buildings to damage. Now many towns, roads, and buildings are built along the coast. The greater number of buildings puts more people in danger. This leads to higher damage totals and a greater risk of injury and death. As a result, modern hurricanes often cause more damage than past hurricanes.

of flooding—especially during the heavy rains and storm surges caused by hurricanes. In 2012 some scientists predicted that climate change could lead to major floods every 3 to 20 years instead of only once a century.

Being Prepared

The more we know about hurricanes, the better prepared we can be. Meteorologists are getting much better at predicting storms. Thanks to satellites and computer maps, scientists understand hurricanes better than ever before. Scientists can track a storm from the moment it forms until it dies out over land. Warnings are issued days in advance, giving people time to prepare.

People follow an evacuation route in North Carolina ahead of 2005's Hurricane Ophelia.

These warnings give people time to evacuate if needed. Towns set up emergency shelters for people whose homes are in the most danger. Many coastal cities have special hurricane evacuation routes that people can follow to safety. People who are far enough inland to safely ride out the storm have time to stock up on food and supplies. They need to be

A homeowner nails boards over his windows to help protect them from hurricane winds and flying debris in preparation for Hurricane Irene.

FURTHER EVIDENCE

Chapter Five discusses climate change and its possible effects on hurricanes. People are sharply divided over the effects of global warming or even whether global warming exists at all. What evidence in Chapter Five supports each point of view? Check out the article on the Web site at the link below. What opinion is stated on this Web site? Write a few sentences about how evidence from the Web site supports a point made in Chapter Five.

Climate Change and Hurricane Sandy

www.mycorelibrary.com/hurricanes

prepared if power goes out for a long time. Many people board up windows as protection from high hurricane winds. Some people stack sandbags around doors and low windows to keep out rising water.

Hurricanes are powerful and destructive storms. Meteorologists are getting better at predicting these storms. But there is no way to prevent a hurricane from forming. The best way to stay safe in a hurricane is to be prepared. And that is something meteorologists can help ordinary citizens do every day.

TEN MIGHTY ATLANTIC HURRICANES

September 1900

Galveston Hurricane: Category 4

This was the deadliest US hurricane in history, killing more than 6,000 people as it roared across Galveston Island in Texas.

September 1926

Great Miami Hurricane: Category 4

This powerful storm hit downtown Miami, Florida. More than 400 people were killed when they went out during the calm eye of the hurricane, only to be trapped when the second half of the storm hit.

September 1928

The Lake Okeechobee Hurricane: Category 4

One of the deadliest hurricanes in history swept through the Virgin Islands and Puerto Rico before striking southern Florida. Approximately 2,500 people died in the storm.

September 1938

Great New England Hurricane: Category 3

This was the worst storm to hit the northeastern United States before Hurricane Sandy. Nicknamed the Long Island Express, the storm raced across Long Island, New York, and continued its destructive path into New England. More than 600 people died.

September 1960

Hurricane Donna: Category 4

Donna first went ashore in Florida, then went out into the Atlantic only to come ashore again in North Carolina. After that, the storm moved up the coast as far as Rhode Island, killing 50 people.

August 1969

Hurricane Camille: Category 5

This storm struck the lower Mississippi River valley in the southeastern United States, killing more than 250 people. Camille's winds were more than 200 miles per hour (320 km/h).

August 1992

Hurricane Andrew: Category 5

This storm affected the Bahamas, southern Florida, and parts of Louisiana. Although the death toll in this storm was low, Andrew destroyed more than 100,000 homes.

August 2005

Hurricane Katrina: Category 3

This hurricane tore across the US Gulf Coast, killing almost 2,000 people. Hurricane Katrina caused more than $100 billion in damage.

September 2008

Hurricane Ike: Category 4

Hurricane Ike raced through the Caribbean, causing extensive damage in Cuba. Then the storm struck Galveston, Texas, with winds of 100 miles per hour (160 km/h). The storm killed more than 100 people and caused $30 billion in damage.

October 2012

Hurricane Sandy: Category 2

The largest hurricane on record to ever hit the United States, Sandy caused more than $50 billion in damage. After hitting the Caribbean, the storm struck the northeastern United States, an area that does not usually see storms this powerful.

STOP AND THINK

Why Do I Care?

You may live in an area of the country that is rarely or never struck by hurricanes. However, your life can still be affected by severe weather conditions. What severe weather conditions are known to strike the place where you live? Do you know how to prepare for severe weather? What steps can you take to be more aware of weather conditions?

You Are There

This book describes what it is like to be in the center of a hurricane. Imagine you are an airplane pilot flying into the eye of a major hurricane. How would you feel? What would you want to experience? Write about what you might see inside the storm.

Say What?

Studying weather can mean learning a lot of new vocabulary. Find five words in this book that you've never heard before. Use a dictionary to find out what they mean. Then write the meanings in your own words and use each word in a sentence.

Surprise Me

This book contains a lot of information about hurricanes and how they develop. After reading the book, what two or three facts about hurricanes surprised you the most? Write a few sentences about each fact. Why did you find them surprising?

GLOSSARY

atmosphere
the mixture of gases that surrounds the earth

barometric pressure
the weight of the atmosphere over an area of the earth's surface

climate
the usual weather in a place over a long period of time

current
the movement of water in an ocean or river

debris
pieces of something that has been broken

diameter
the distance from one side of a circle to the other through its center

evacuate
to move away from an area because of danger

eye
the center of a tropical storm or hurricane that has low pressure, light winds, and little or no rain

eye wall
an area of storm clouds that surround the eye of a hurricane and contain severe weather

meteorologist
a scientist who studies the weather

satellite
a spacecraft that orbits Earth and sends back information

LEARN MORE

Books

Bailey, Rachel. *Superstorm Sandy.* Minneapolis: ABDO, 2014.

Fradin, Judith Bloom, and Dennis Brindell Fradin. *Hurricane Katrina.* New York: Marshall Cavendish, 2010.

Paul, Gill. *See-Through Storms.* Philadelphia: Running Press Kids, 2006.

Web Links

To learn more about hurricanes, visit ABDO Publishing Company online at **www.abdopublishing.com**. Web sites about hurricanes are featured on our Book Links page. These links are routinely monitored and updated to provide the most current information available.

Visit **www.mycorelibrary.com** for free additional tools for teachers and students.

INDEX

ABOUT THE AUTHOR

Joanne Mattern has always been fascinated by wild weather. She is the author of many books for children. Joanne lives in New York State with her husband, four children, and many different pets.